Day and Night

Jen Green

PowerKiDS
press.
New York

Published in 2008 by The Rosen Publishing Group, Inc.
29 East 21st Street, New York, NY 10010

Copyright © 2008 Wayland/The Rosen Publishing Group, Inc.

First Edition

Picture credits:
Cover Digital Vision, 1 and 16 NASA, 4 Digital Vision, 5 Dreamstime.com,
6 Dreamstime.com/Goce Risteski, 7 Dreamstime.com/Edyta Pawlowska,
8 NASA, 9 Corbis/Rolf Brudererß, 10 Digital Vision, 11 Dreamstime.com/Elaine
Davis, 12 Dreamstime.com/Laura Bulau, 13 Dreamstime.com/Patrick Hermans,
14 Dreamstime.com/Vladimir Pomortsev, 15 Alamy/Dynamic Graphics Group
/i2i, 17 Dreamstime.com/Ben Goode, 18 Dreamstime.com/Jack Dagley,
19 Digital Vision, 20 Dreamstime.com/Bonnie Jacobs, 21 Dreamstime.com/Jessi
Eldora Roberston

Produced by Tall Tree Ltd.
Editor: Jon Richards
Designer: Ben Ruocco
Consultant: John Williams

Library of Congress Cataloging-in-Publication Data

Green, Jen.
  Day and night / Jen Green. — 1st ed.
      p. cm. — (Our Earth)
  Includes index.
  ISBN-13: 978-1-4042-4275-3 (library binding)
 1. Earth—Rotation—Juvenile literature. 2. Day—Juvenile literature. 3. Night—
Juvenile literature. I. Title.
  QB633.G697 2008
  525'.35—dc22
                                    2007032591

Manufactured in China

Contents

Day and night

Day and night are one of nature's **cycles**. At **dawn**, the Sun rises, and it gets light. The Sun **sets** at **dusk**, then it is night. After night, a new day begins. This cycle never changes.

❯❯ During the day, the place where you live is lit by the Sun and you can see easily.

▲ Nighttime is dark and we use electric lights to see.

During the day, the Sun provides light, which helps us to see. Most people are busy by day, and sleep when it is dark. Some animals do the same, but others sleep during the day and are active at night.

Fact

Living things depend on light and heat from the Sun for life.

What causes day and night?

The Earth is a rocky planet that spins around. As one side spins into the sunlight, day begins. As it spins away from sunlight, it moves into night. The Earth takes 24 hours to spin once. This is one day.

The part of the Earth facing the Sun has daytime, and at the same time, the part facing away has nighttime.

❯❯ This is a 12-hour clock. The hour hand completes two circles in one day.

The Earth moves around the Sun on a journey called an **orbit**. The Earth takes just over 365 days to travel around the Sun. This is called a year.

Our star, the Sun

The Sun is a huge, fiery ball of burning gases. It is a **star**, like the stars we see in the night sky. The **temperature** on the Sun's surface is 9,900°F (5,500°C), and the inside is much hotter.

⊗ The Sun is more than 600,000 miles (1 million km) wide.

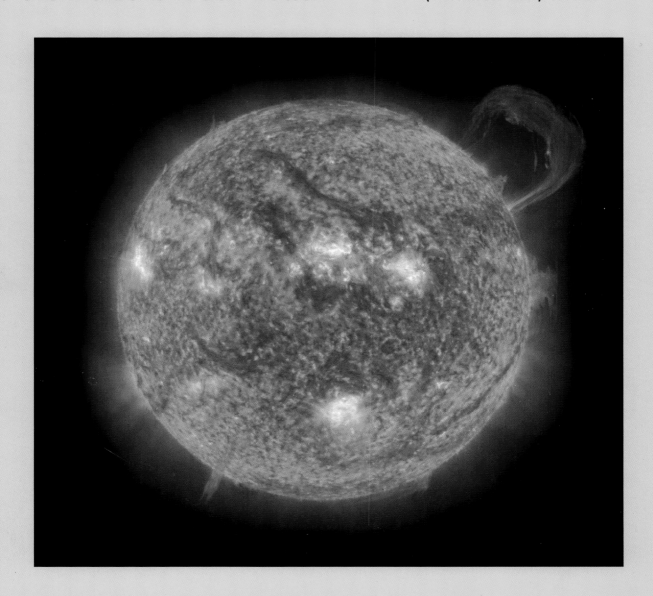

As the Sun's gases burn, **rays** stream out in all directions. The Sun's rays travel in straight lines. This gives us sunshine. Shadows appear as objects, such as a tree, block the rays of sunlight.

Fact

The Earth is about 91 million miles (147 million km) from the Sun.

Protect your eyes by wearing sunglasses in bright light. Never look straight at the Sun.

9

Dawn

As the Earth spins around, part of its surface moves into the sunlight. The Sun rises in the east each morning. This is called dawn. The Sun is not actually moving—the Earth is spinning around.

Dawn marks the end of the night and the beginning of the day.

A butterfly's body temperature is the same as its surroundings. It needs to sit in the morning Sun to warm up before it can fly around.

Some animals sit in the sunlight to warm their bodies before they get moving. People who work outdoors, such as farmers, may also start work at dawn.

Morning and noon

During the morning, the Sun climbs in the sky and moves to the west. As the Sun gets higher in the sky, the temperature rises.

❯❯ Many flowers open their petals and turn toward the Sun in the morning.

Fact

Plants use sunlight to make their food. They cannot grow without light.

▶▶ Daytime hunters, such as eagles, need light to see to catch their food.

As the day warms up, animals become more active, looking for food to eat. At noon, the Sun is at its highest point in the sky.

Afternoon and evening

Early afternoon is often the warmest time of the day. In hot places around the world, people and animals rest in the shade to avoid the heat of the Sun.

Desert animals, such as these bat-eared foxes, keep cool by sitting in the shade.

⯅ Shadows are longer in the evening, because the Sun is low in the sky.

As the Sun **sets** in the west, the air gets cooler. Flowers close their petals. Some animals return to their homes, and many people get ready to go to bed.

Nighttime

The Moon and stars shine at night. They shine by day, too, but the sky is too bright to see them. Many stars are bigger than the Sun, but they look small because they are so far away.

◆ Stars shine in lots of different colors, including white and blue, as this close-up of a pair of stars shows.

▶ Every month, we see different amounts of the Moon, from a thin **crescent** to a whole disk. These are called phases of the Moon.

The Moon is a rocky ball that goes around the Earth, just as the Earth goes around the Sun. The Moon's rocky surface shines because it **reflects** the Sun's light. The Moon travels around the Earth once every 28 days.

Fact

The Moon is about 238,855 miles (384,400 km) from the Earth.

Awake at night

It makes sense for us to sleep at night. We need to rest and it is dark, so we cannot see anything. However, many people, including police officers, firefighters, and nurses, may have to work at night.

Firefighters tackle a blaze at night.

▶▶ Animals such as this owl sleep during the day and are awake at night.

Some animals wake up at dusk and go hunting. Bats, mice, and owls are all **nocturnal**. These animals have large eyes that see well in dim light.

Fact

*Bats track moths at night by squeaking and listening for **echoes** that bounce back off of the moths.*

What time is it?

When it is day, the other side of the world has night. Around the world, noon is when the Sun is highest. Noon comes later the farther west you live, so clocks in different countries tell different times.

⬇ When it is noon in London, most people are awake and children are at school.

At the same moment, it is four o'clock in the morning in Los Angeles, and most people are asleep.

We are used to living in our own **time zone**. When you fly to a different country, your body has to adjust to the new time, and you may feel tired for a while. This is called **jet lag**.

Activities

Making a Sun clock

The Sun appears to move across the sky at a constant speed. You can use this speed to make your own shadow clock.

WHAT YOU NEED

- **A sunny day**
- **Level ground**
- **A long stick**
- **Pebbles**
- **A clock**

1. Push the stick into the ground. At the start of an hour, mark where the stick's shadow points to using a pebble.

2. Wait another hour and mark where the stick's shadow points to again. Continue marking the shadows every hour to complete your Sun clock.

Use the stars to find north and south

You can use the stars in the night sky to find out which way is north and south. All you need is a clear view of the night sky and an adult to keep you safe.

Northern half of the Earth

Pole star

Big Dipper

B

A

1. If you live in the northern half of the Earth, find the constellation called the Big Dipper.
2. Draw an imaginary line between the two end stars (marked "A" and "B"), and follow the line until you come to another star. This is the Pole Star and it points to the north.

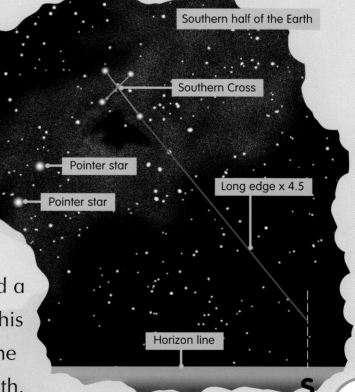

Southern half of the Earth

Southern Cross

Pointer star

Pointer star

Long edge x 4.5

Horizon line

1. If you live in the southern half of the Earth, find the Southern Cross.
2. Draw a line between the two stars on the longest part of the cross.
3. Extend this line about four and a half times the cross's length. This guides you to a point above the **horizon**, which is the way south.

Glossary

Crescent The thin, bent shape that the Moon shows during its phases.

Cycles Repeating patterns of events.

Dawn The start of the day when the Sun appears.

Dusk The end of the day, when the Sun sets.

Echoes When sounds repeat after bouncing off of an object.

Equator An imaginary line around the Earth.

Horizon Where the ground meets the sky.

Jet lag When you feel tired after a long trip.

Nocturnal Animals that are active at night.

Orbit The curving path followed by the Earth as it moves around the Sun.

Reflects When light rays bounce off of a surface.

Rays Beams of light.

Sets When the Sun sinks toward the horizon.

Star An enormous ball of burning gas. There are millions of stars in the night sky. Our Sun is a star.

Temperature How hot or cold something is.

Time zone An area where the same time is used.

Index

Web Sites
Due to the changing nature of Internet links, PowerKids Press has developed an online list of Web sites related to the subject of this book. This site is regularly updated. Please use this link to access this list:
www.powerkidslinks.com/earth/dayni